Kona Village Apartments Fire
Bremerton, Washington

Investigated by: John Kimball

This is Report 121 of the Major Fires Investigation Project conducted by Varley-Campbell and Associates, Inc./TriData Corporation under contract EME-97-CO-0506 to the United States Fire Administration, Federal Emergency Management Agency.

Department of Homeland Security
United States Fire Administration
National Fire Data Center

U.S. Fire Administration Fire Investigations Program

The U.S. Fire Administration develops reports on selected major fires throughout the country. The fires usually involve multiple deaths or a large loss of property. But the primary criterion for deciding to do a report is whether it will result in significant "lessons learned." In some cases these lessons bring to light new knowledge about fire--the effect of building construction or contents, human behavior in fire, etc. In other cases, the lessons are not new but are serious enough to highlight once again, with yet another fire tragedy report. In some cases, special reports are developed to discuss events, drills, or new technologies which are of interest to the fire service.

The reports are sent to fire magazines and are distributed at National and Regional fire meetings. The International Association of Fire Chiefs assists the USFA in disseminating the findings throughout the fire service. On a continuing basis the reports are available on request from the USFA; announcements of their availability are published widely in fire journals and newsletters.

This body of work provides detailed information on the nature of the fire problem for policymakers who must decide on allocations of resources between fire and other pressing problems, and within the fire service to improve codes and code enforcement, training, public fire education, building technology, and other related areas.

The Fire Administration, which has no regulatory authority, sends an experienced fire investigator into a community after a major incident only after having conferred with the local fire authorities to insure that the assistance and presence of the USFA would be supportive and would in no way interfere with any review of the incident they are themselves conducting. The intent is not to arrive during the event or even immediately after, but rather after the dust settles, so that a complete and objective review of all the important aspects of the incident can be made. Local authorities review the USFA's report while it is in draft. The USFA investigator or team is available to local authorities should they wish to request technical assistance for their own investigation.

For additional copies of this report write to the U.S. Fire Administration, 16825 South Seton Avenue, Emmitsburg, Maryland 21727. The report is available on the Administration's Web site at http://www.usfa.dhs.gov/

U.S. Fire Administration

Mission Statement

As an entity of the Department of Homeland Security, the mission of the USFA is to reduce life and economic losses due to fire and related emergencies, through leadership, advocacy, coordination, and support. We serve the Nation independently, in coordination with other Federal agencies, and in partnership with fire protection and emergency service communities. With a commitment to excellence, we provide public education, training, technology, and data initiatives.

ACKNOWLEDGMENTS

The United States Fire administration gratefully acknowledges the assistance, cooperation, and professionalism of the men and women of the Bremerton Fire Department under the direction of Chief Allison Duke, III.

Photos and tactical diagrams courtesy of the Bremerton Fire Department.

News Photo and diagram on page ?? courtesy of the Bremerton Sun.

Reviewers: Michael Tamillow
Battalion Chief
Fairfax County Fire and Rescue Department
Fairfax, VA

Douglas Stutz, Ph. D.
Director, Fire Science Program
Miami-Dade Community College
Miami, FL

TABLE OF CONTENTS

Kona Village Apartments Fire
Bremerton, Washington
November 1997

EXECUTIVE SUMMARY

On November 13, 1997, an early morning fire in a large apartment building in Bremerton, WA, caused the death of four elderly residents and forced the evacuation of an additional 150 residents. Estimated property damage was in excess of $7.5 million.

The fire, determined to be accidental, started in an unoccupied apartment (Apartment 316) and spread rapidly, trapping many residents. At least 21 people had to be rescued by firefighters using ground ladders. The construction features of the non-sprinkled building contributed to the fire spread. The blaze quickly involved the exterior walkway in the courtyard from the apartment door, communicated upward via pipe chases and utility shafts, and extended from the exterior rear window of the apartment of origin. Once in the attic, the inadequate and poorly constructed fire walls allowed rapid horizontal fire spread.

Fire units arrived in less than a minute and a half and all efforts were concentrated on rescue for the first 20 minutes of response. Firefighters were hampered by lack of hydrants within the complex, lack of access to the apartment entrances, and lack of sufficient area in which to operate. Three full alarms and three strike teams were needed to control the fire, which was declared at 0745. Residents were relocated via city buses to a nearby church. In addition to the efforts of many local fire departments, various agencies of the Bremerton City government and private relief groups worked together to coordinate logistical support in the suppression, operation, and relocation and relief efforts. The Federal Bureau of Alcohol, Tobacco, and Firearms (ATF) Task Force assisted Bremerton fire officials in determining the origin and cause of the fire.

The apartment building was constructed in 1971 without sprinkler and central alarm systems. (A retrofit ordinance requiring sprinklers was not in effect at the time of the fire.) City fire officials had written the building owners at least twice requesting that they install improvements in built-in fire protection, including fire sprinklers.

This fire provided the impetus for the city to initiate an amendment to its building code to require sprinkler protection for such occupancies. The fire department has begun a risk prioritization program in which occupancies are classified as to height, occupancy load, specialty use, square footage, and unique access or egress problems. Moreover, a bill has been reintroduced in the Washington State legislature to require sprinkler retrofits for multiple occupancy buildings, in part due to the publicity from this fire.

INCIDENT NARRATIVE

Early on Thursday morning, November 13, 1997, Bremerton fire units responded to a serious vehicle accident that required helicopter transport of the injured. The crew of Engine 3 had performed standby duty for the medevac helicopter and had reported at 0557 that they were leaving the scene and returning to their station. Their usual route of travel took them past the Kona Village apartment building, where they noticed nothing unusual. This large square-shaped structure of interconnected apartment wings and an open interior courtyard was well known to Bremerton firefighters because of it status as a target hazard. It was a large wood-frame building housing almost 200 people, many of them elderly living on the upper floors. The building was a closed square shape with apartment entrances facing an interior courtyard with no vehicular access to the courtyard.

As the crew of Engine 3 was fueling the vehicle at the station, an alarm was dispatched for Kona Village Apartments at 1717 Sheridan Road reporting fire in Apartment 316. The time of dispatch was 0608.

When Engine 3's crew made the one-minute-and-ten-second response, they reported on the scene with heavy smoke showing and quickly requested a second alarm. They reported to the west side of the building and used the closest access walkway to Apartment 316.

As the lieutenant of Engine 3 stretched a 1-3/4-inch pre-connected line to the center court, the smoke increased in intensity and flashed into visible flames from the third and fourth floor. Simultaneously, the driver of Engine 3 witnessed fire flash out of the windows on the exterior of the courtyard. Silhouetted by heavy smoke, occupants immediately appeared from windows of the apartments screaming for help. The lieutenant requested a third alarm. The pump operator gave a quick situation report of people trapped and reported help needed for rescue. (Bremerton staffing is two people each for engine and medic units.)

The lieutenant of Engine 3 operated his preconnected line to attempt to protect the escape of occupants from the interior walkway and stairs. Residents were rousing other residents by knocking on doors, and escaped by helping each other through the smoke. Other Bremerton engines and medic units arrived, as did second alarm units from nearby departments. For the first 20 minutes all efforts were directed toward the rescue of occupants. The line to the interior courtyard went dry as Engine 3's booster tank was depleted, leaving firefighting crews operating in the interior courtyard without cover. On the outside of the complex, the efforts of four engines and one ladder truck were focused on the deployment of ladders and removal of tenants. The age and infirmity of many of the occupants compounded the rescue effort.

Incident command procedures were implemented and efforts transitioned from rescue to fire containment with secondary search and rescue. Exterior master streams were deployed, including at least two portable monitors in the courtyard as well as numerous 2-1/2-inch handlines. Master streams, including a ladder pipe, were deployed at the southeast corner of the building in an attempt to cut off the spread of fire. This was generally successful, but embers dropping down from above caused some fire extension on the east side.

Control of the fire was established at approximately 0745 with extensive overhaul continuing for several days. Bremerton fire investigators were assisted by an ATF task force consisting of more than 30 people who worked on site for three days.

The rapid spread of the fire was mainly due to the extensive wood frame construction and sub-standard fire stopping. This included not only the attic space, but also wooden interior walkways that trapped superheated gases from burning apartments and channeled them horizontally. Vertical openings in the interior of the apartment in the form of pipe channels and wiring shafts were also a factor. The buildings were neither sprinkled nor equipped with an interconnected alarm system. Firefighters were also hampered by the lack of access to the interior courtyard, insufficient clearances for vehicles on the outside of the building, insufficient hydrants in the complex, obstructed hydrant access, and poor available water flow.

SUMMARY OF KEY ISSUES

Issue	Comments
1. Building Construction	Lack of sprinklers
	Substandard fire stopping in attic
	Substandard fire stopping in walls
	Wooden stairwells and walkways
	Lack of adequate firefighter access to apartment entrances
	Lack of interconnected smoke detectors and exterior alarm
	Insufficient number and spacing of hydrants
	Accessibility of hydrants
	Lack of sufficient operating space for fire vehicles
2. Fire Operations	Insufficient staffing of first alarm units
	Failure to establish a water supply to first engine
	Lack of aerial device on first alarm
3. Code and Regulatory Issues	Comments
	Retrofit ordinances
	Applicability of building code

BUILDING FEATURES

Construction

The Kona Village apartment building was completed in 1971 of frame construction with a low-angle pitched roof with composition shingles. The outside dimensions of the structure were 270 feet by 320 feet. The building was almost completely constructed of wood; floor, roof, and structural members. The foundation was concrete. A protective coating of asbestos was sprayed on the ceiling of the kitchens and the garage areas. (The discovery of asbestos was made during the overhaul stage and presented a troublesome challenge to the fire department.) The address front, on the north face, consisted of a two-story portion with a 20-foot-wide center walkway providing access to the interior courtyard. The rental office was located in the first apartment on the ground floor just to the right of the walkway. The courtyard included a swimming pool and associated buildings.

The rest of the building (east, south, and west wings) consisted of three stories of wood-framed construction (built over a first floor of concrete block) joined to form a square. The ground floor of each unit in this section contained occupant parking units, laundry rooms, and storage rooms. The upper three floors were apartment units of identical configuration stacked vertically. The kitchen/bathroom utility shafts abutted one another on each floor. Access to the interior courtyard from the parking lost consisted of open walkways or "breezeways" approximately six feet wide. There were ten of these breezeways spaced around the three four-story wings.

Exit paths

The exits consisted of six-foot walkways from the apartment entrances around the circumference of the interior courtyard leading to stairs. The walkways were made of wooden 3-by-8-inch tongue-in-groove boards over 4-by-12-inch beams covered with indoor/outdoor carpeting. Several firefighters reported seeing this carpeting burning vigorously.

Access to the second and third floors was by means of a combination of open stairways, closed stairwells, and elevators. One open stairway was located at each corner of the wings and four were spaced on the north block.

The partially closed stairwells extended from ground to roof and had 12-by-8-foot interior dimensions.

A total of four closed stairwells served the east, south, and west wings. All of the stairways were totally of wooden construction. The elevators were built on the east and west blocks near the corner of the south block. The elevator enclosures were wood-framed construction boxed around the elevator structure and machinery.

Fire Protection Systems

The building was equipped with smoke detectors in individual units that were not linked to a building system. There was no master alarm/pull station or central alarm system.

There were no sprinklers or other active fire suppression systems. Fire extinguishers were located in boxes on each floor. Six standpipe outlets were located in the west, east and south blocks with 100 feet of 1-1/2-inch hose and nozzle in each. These were piped to the domestic water system via a 1-inch line with no fire department supply connection. (At least two of these were used by residents at the fire, but were obviously ineffective at suppressing the blaze.)

Building Code

The building was constructed under the 1967 building code, the 21st edition of the Life Safety Code of the Uniform Building Code.

The complex was occupied in June 1971 and the complex had been in operation continuously since then with no major renovations.

Inspection History and Violations

After a fire in 1986 that identified shortcomings in the fire stopping in the attic, the city fire marshal submitted a letter to the building owners. The letter expressed concerns over the condition of the building, especially the lack of sufficient fire stopping and alarm system.

In 1994 the city sent another letter to the building owners requesting consideration of fire sprinklers in the Kona Village apartment building. The owner reportedly verbally refused to consider this request, and the city believed it did not have the authority to enforce such a request. The issue was not pursued.

The building had been inspected within the last two years with no major code violations or infractions. It was due for routine fire inspection during the month of November 1997.

Occupancy

The Kona Village Apartments attracted a variety of residents, but included a high concentration of elderly people on pensions or other fixed income. A substantial number of these people had limited mobility and required prosthetic devices or aids to walk. One of the rescuers spoke of people exiting "in their walkers, crutches, and wheelchairs" from the upper floors. Most of these residents had been at the Kona Village project for many years.

Many of the tenants, however, were younger couples employed in heavy industry, such as the nearby ship repair and support industry. Other residents were employed by the armed forces especially the U. S. Navy, which has extensive presence in the area.

Social Environment

The Kona Village complex was a very desirable rental location due to three main factors. First was the relatively low rental cost per month. The age of the complex was a factor permitting this reasonable and stable rent over the years. Kona Village was an especially attractive location for those on fixed incomes. The second factor was the building's reputation as a well-maintained occupancy. It was known in the area as a safe, clean place to live. The tenure of many of the occupants served to perpetuate a sense of community and security. The third factor was the picturesque view the apartments afforded by being on the highest point in the area. To the south was Mt. Rainier and to the west were the Olympic Mountains. The area has one of the most desirable vistas in the west Puget Basin area.

Units on the higher floors had the more desirable view and were acquired on a seniority basis. This practice placed more of the elderly residents on the upper floors. The building was situated in a working class neighborhood of single family homes at the intersection of two major residential streets. A church, a school, the city water authority office, and Bremerton Fire Station 3 were all within 3000 feet of the complex.

Diagram 1 shows the features of the Kona Village complex. Note the U-shaped construction of three living floors built over the parking/storage/utility floor. The north side is the two-story portion, which housed the rental office.

BACKGROUND

Bremerton is an independent city within the boundaries of Kitsap County on the western fringes of the Seattle-Tacoma metropolitan area. The city was settled in 1891 after establishment of the Puget Sound Naval Shipyard. Industry in the area centers around ship repair and the maintenance and associated support functions for naval operations. Several military installations in the area also support employment for the population. The population of Bremerton is approximately 40,000.

The city government operates with an elected mayor and city council. The fire chief reports directly to the mayor. Kitsap County borders the city and provides various support functions, including disaster assistance and logistics from the Department of Emergency Management. The authority having environmental jurisdiction is the Puget Sound Air Pollution and Control Agency (PSAPCA.)

The Bremerton Fire Department (BFD) provides fire and emergency services through a 52 member department under the direction of a fire chief. The department is organized into the separate functions of Prevention & Investigations, Training, and Medical, each with a captain as officer-in-charge. The operations group is organized with a captain serving a shift command on each of the three shifts. All of the functions report directly to an assistant chief, who reports to the fire chief.

The suppression forces are deployed from three stations, each with an engine and two with a medic unit. An aerial ladder is normally located at Fire Station 1, but was out of service at the time of the Kona Village fire. Assigned staffing consists of two people each on medic units and the engines. The engine officer is a lieutenant, with Station 1 having a third member assigned to the engine and the engine officer acting as shift commander at the captain rank.

Operations were governed by a set of standard operating procedures. Of particular importance is SOP #2-22, Incident Command Procedures. These procedures are based on Nationally recognized practices and have served the BFD well. The command protocol is clearly delineated; the first arriving officer has command until relieved by a more senior officer. Typical operations have the first-arriving officer retaining command until the incident is controlled or the first-arriving officer is relieved by the next-in officer. Generally the shift commander from Fire Station 1 would relieve the first-in officer if appropriate. A staff duty officer (chief or assistant chief) is notified by radio page and has the option of responding and assuming command. (On the Kona Village incident, the fire chief was notified and was on the scene within five minutes of the initial report of the first engine.)

Attack procedures call for the first engine to report directly to the fire location and initiate one of the three command options:

1. Nothing Showing Mode; investigate further

2. Fast Attack Mode

    A. Need for offensive fire attack

    B. Need for life safety (i.e., critical rescue must be achieved in a compressed time)

    C. An incident where the safety and welfare of the firefighters is a major concern

    D. Obvious working incidents that require further investigation by the company officer

3. Command Mode

Applied in situations that require strong command by virtue of the size of the fire, the complexity or potential of the occupancy, or the possibility of extension. Such situations require strong, direct overall command from the outset. In these cases, the company officer will initially assume a command position and maintain that position until relieved by a command officer.

Typical operations in the BFD call for the first engine to proceed directly to the fire area without laying a supply line, and to attack the fire from the booster tank while the next available engine establishes water supply. Additional attack and back-up personnel are deployed as they become available. The first engine establishes and maintains command until relieved by a senior officer.

FIRE: ORIGIN AND SPREAD

Location, Detection, and Reporting

The fire was first discovered by the resident manager while he was on his morning rounds delivering newspapers, just before 0600. He heard a smoke detector sounding on the third floor in the west wing area and gained access to the suspected Apartment 316. The smoke and heat forced him to crawl on the floor. Just inside the apartment door he saw evidence of fire in the area in the vicinity of the bedroom. The resident manager determined that there was no one in the apartment and left (possibly leaving the apartment door to the walkway open) to place a 9-1-1 call from the rental office. The time was then 0606. He returned to the apartment to attempt to fight the fire. The process from his initial discovery of the fire to placing the 9-1-1 call took about six minutes. During those few minutes, the fire rapidly grew in intensity and he was forced to retreat. A resident in Apartment 315 also placed a 9-1-1 call and described the rapidly deteriorating conditions to the dispatch center. The dispatcher directed her to the apartment balcony (where she was rescued via ladder by Bremerton fire units just as the balcony flashed into open flame.) Bremerton fire units were immediately dispatched and arrived on the scene shortly thereafter.

Alarm and Alert

Occupants were alerted by other occupants and firefighters knocking on doors. Some were awakened by the odor of smoke and the noise of firefighting operations. At least one of the first alarm units sounded their siren and air horn in a general alarm effort to awaken all tenants. In the adjacent blocks on the north and east sides, many residents were still asleep well into the progress of the fire. They had to be awakened and assisted out by firefighters.

Fire Spread

The fire spread rapidly from Apartment 316. The smoke darkened down and flashed out of the front door. The overhanging fascia board from the walkway above served to trap heated gases and smoke as they progressed out from the apartment. As the fire in the apartment grew in intensity, the heat was trapped under the wooden exterior walkway.   The trapped gases increased in temperature, pre-heated the wooden walkway joists and framing, and flashed over. (The first responding lieutenant clearly described the process of the smoke from the apartment as intensifying, darkening into "flashover smoke," and igniting into free burning fire.) This allowed the fire to spread horizontally on the third floor. Because of the wooden construction and the combustible barriers, the fire in the walkway spread upward rapidly and continued the process on the fourth floor. As the fire originated in an apartment near the juncture of west and south sides, this phenomenon allowed rapid fire spread vertically and horizontally in the two wings on the interior courtyard side.

At the same time, the fire was progressing upward via interior openings in pipe chases and utility shafts. These utility shafts serviced the bathroom and kitchen area and provided channels for heat, smoke, and gases vertically to the next floor and into the attic area. There was only one layer of drywall (1-hour rating, ½ inch) on the ceiling of each unit; this also allowed fire to spread to the fourth floor units. Once the fire entered the attic, it spread very rapidly. Fire stops were reported to have been placed between every four units. These stops were of uncertain integrity and in any case did not provide full firewall-type protection.

The fire was also progressing via exterior vertical flame spread on the outside wall. Once the window broke in Apartment 316, fire rapidly shot out and moved up the combustible wood exterior to involve the fourth floor and roof area.

In the later stages of the fire, as fire units sought to contain the spread by concentrating on the attic, sparks and embers dropped down from the fourth floor and attic and ignited first and second floor apartments on south and east sides.

At least 21 occupants were rescued via exterior ladders on the outside wall. All people escaped with only the clothes on their backs.

Photo 1 was taken at the southeast corner of building, shows fire conditions twenty minutes after dispatch.

The area of origin and location of the fatalities is at the left in the photo. The fire has extended horizontally through the attic area and dropped down via pipe chases and utility shafts.

Photo 2 shows interior courtyard operations. Crews are deploying a portable monitor and multiple handlines to cutoff exterior fire extension. The two-story portion (north) is visible at the right. The enclosure to the left is the elevator. The apartment of origin is to the upper left.

RESPONSE OPERATIONS

Fire Department Dispatch

The first alarm was placed at 0608 for a fire at 1717 Sheridan Road in Apartment 316. Bremerton Engines 3, 1, and 2 Puget Sound Shipyard Engine 15 (Telesquirt), and Medic 3 were dispatched. Engine 3 requested a second alarm as they marked on the scene with heavy smoke showing. Travel time was approximately 1 minute and 10 seconds.

Initial Operations; Rescue and Attack

Engine 3, consisting of the driver and lieutenant, proceeded directly into the complex via the parking lot on the west side. Consulting the pre-plan en route, they took a vantage point at the closest access breezeway to the interior courtyard. The lieutenant stretched a 1 3/4–inch attack line to the center courtyard with the idea of cutting off the fire spread and providing water stream protection and cover for escaping residents. As the line was being charged, the windows in the rear blew out and visible fire was reported from the front and rear. Heavy fire developed in the third and fourth floor on the courtyard side.

The operator of Engine 3 reported occupants appearing at many windows in the rear surrounded by heavy smoke coming from the inside their apartments. She immediately broadcast a situation report of a severe rescue problem and requested help on the west side. All efforts were directed to deployment of ground ladders for rescue of the trapped occupants. As more units arrived they were directed to the exterior of the west and south sides to assist in the rescue effort. Engine 1 assumed command as per BFD procedure and established the command post on the address front at the corner of the north and west sides. The operator stretched a 5-inch supply line to the hydrant and deployed a stationary deluge gun from the engine over the front roof (side A) into the court yard. The goal was to provide water cover for the occupants and firefighters in the courtyard. **For the first ten minutes of the fire, no other water was flowed except the 1-3/4-inch line from Engine 3 to the interior courtyard and Engine 1's deck gun.**

As Engine 3's booster tank was depleted, the lieutenant's position in the courtyard became untenable. He requested a supply line for Engine 3 to cover rescues. Engine 2 was directed to stretch a supply line to Engine 3. This provided difficult because of the close clearance in the parking area in the rear, as well as the fact that residents were attempting to leave in their vehicles. At the same time the third and fourth floor partially collapsed at the intersection of the west and south sides, carrying the body of one of the fire victims into the parking lot and showering fire personnel with debris. Deteriorating conditions due to the radiant heat and potential for further collapse caused Engine 3 to relocate to the rear of the south side. This also complicated efforts by Engine 2 to establish a supply for Engine 3.

The Bremerton Fire Chief responded from home and assumed command at approximately 0625. All efforts continued to focus on rescue until about 0640.

Diagram 2 indicates origin of fire and position of first arriving engine. Heavy smoke was evident on the west side and coming from the courtyard.

Diagram 3 indicates the position of second arriving engine (E-1) an extent of rapid fire travel. Conditions deteriorated from heavy smoke to heavy fire in a matter of seconds. E-1 hooked directly to hydrant and directed a large-caliber deck gun stream over the roof in an attempt to cut the radiant heat and protect firefighters and residents in the courtyard.

Second Phase of Operations

About 20 minutes after arrival of the first units, the fire strategy was shifted from the rescue phase to containing the fire. A more extensive command structure was established and passport accountability was implemented. The fire chief designated an operations officer, safety officer, and interior division. Command officers from mutual aid departments were also assigned liaison functions by the incident commander. The strategy was to hold the fire to the south and west blocks and cut off the spread at the SE and NW corners. Hand lines were deployed to the interior. At first these were 2-1/2-inch fog nozzles, but the need for the reach and penetration of solid streams was quickly recognized and implemented. At least two portable monitor devices were placed in service in the courtyard and used to good effect. The ladder pipe of Truck 51 was instrumental in cutting off the spread on the SW corner. The combination of water penetration by exterior master streams and the mobility of hand lines halted the spread of the fire. However, fire was dropping from the attic and fourth floor into lower floors.

Third Phase

The third phase of operations consisted of extinguishing efforts after the fire spread had been checked. Though the rapid spread of fire was stopped, extensive firefighting continued as individual rooms and apartments became involved in fire. Tedious room-to-room attack, extinguishing, and overhaul operations continued for several hours.

Early in the incident, crews were dispatched to mark the exterior sides according to BFD incident command protocol. This served to provide consistency in nomenclature as to building sides. As operations progressed, the paint was used to designate the areas that had been cleared of occupants by primary and secondary searches.

Units on the initial three alarms were supplemented by two engine strike forces (three four-person engines with a command officer) and one ladder strike force (three ladder units with a command officer). These resources were utilized by command to rotate, reassign, and relieve personnel. In the second and third phase, the tactical objective was to place two attack crews on each floor with a four-engine rapid intervention force available at all times.

Diagrams 4 and 5 indicate strategy in the third phase. Extinguishing required a combination of portable master streams and handlines in the interior court and ladder pipes from the outer perimeter.

Sprayed-on asbestos in the kitchens and garages was discovered during the early overhaul phase. Notification was made to PSAPCA and the storm drain intakes were sealed with plastic sheeting. Environmental authorities took samples to test for asbestos levels. Residual water in the storm drain sumps was pumped out by properly equipped vacuum trucks. All responding companies were notified of the presence of asbestos and directed to properly decontaminate all gear, clothing, and equipment. The asbestos also complicated efforts of residents to reoccupy their apartments and retrieve their belongings. This became more than an inconvenience, because prescription medicine, cash, and valuable items had been left in the apartments.

The Bremerton/Kitsap County mutual aid area also is served by a Public Information Officers Task Force. This group consists of PIOs from the area departments who respond to large incidents in neighboring jurisdictions to assist the home jurisdiction. This Task Force proved to be quite valuable in the Kona Village fire, because the need for PIO services was far greater than the BFD could handle alone.

The on-duty shift was relieved at 0930 on the fireground. All first alarm units were invited to a Critical Incident Stress Debriefing the next morning. Nationally recognized protocols for CISD were in effect. Separate debriefings were conducted as appropriate, as were after-action critiques.

The south block was totally destroyed by fire, the west block was 60 percent destroyed, and the east block was about 30 percent destroyed. Overhaul continued for several days concurrent with the search for victims. The cause and origin investigation continued as well.

Logistics

All requests for logistics were channeled from the command post, either invoking pre-planned procedures through Central Communications or from City of Bremerton officials. Support in the form of food, rehabilitation units, portable comfort stations, breathing air, lights, and power generators was required. Kitsap Transit provided a bus at 0700 for shelter and transport of victims. The nearby church opened its doors for a relief and relocation shelter. Red Cross, Bremerton City, and Kitsap County Emergency Management officials set up a shelter at that location.

Interagency coordination

Pre-determined procedures and good working relationships made interagency cooperation excellent. One reason for this was that in June 1997, the management staff of the City of Bremerton, Kitsap County, and four other area cities attended the Disaster Preparedness Training Session for city officials at the Emergency Management Institute (EMI) at the National Emergency Training Center at Emmitsburg, MD. The fire chief directly credited this class as having a positive effect on the outcome of the level of emergency coordination at the Kona Village fire. Agencies who operated or provided support are listed in Appendix A.

CASUALTIES AND PROPERTY LOSSES

Deaths

Four people were killed in the Kona Village fire. They were 75, 76, 78, and 91 years of age. One resided in Apartment 417, two in 419, and one in 423. All deaths were due to smoke inhalation.

Injuries

Twelve people, eleven civilians and one firefighter, were injured in the fire. The most serious of these was a broken ankle. The summary of injuries is:

> 3 – Smoke inhalation
>
> 1 – Burns to the hands, smoke inhalation
>
> 1 – Burns to the hands
>
> 1 – Fractured ankle
>
> 2 – Leg pain
>
> 1 – Foot pain
>
> 1 – Leg laceration
>
> 1 – Unspecified injury
>
> 1 – Anxiety attack
>
> None of the injuries required hospitalization.

The injuries to the extremities (5) were caused by people jumping to escape the flames and smoke. These and the number of burn and smoke injuries (9) attest to the rapidity of the fire spread.

The average age of the injured was 55 years.

The average age of the fatalities was 80 years.

All of those who died resided on the fourth floor.

Distribution of injuries: Fourth floor – 6

> Third floor – 3
>
> Second floor – 2
>
> Firefighter – 1

Ten of the casualties were from the fourth floor, representing 63 percent of all killed or injured.

Property damage

The fire caused $7.5 million in property damages to the building, contents, and automobiles. At least three properties on the south side suffered minor damage such as melted siding and damage to autos.

Indirect damage will include sample testing, site decontamination, and remediation of the asbestos, as well as efforts to contain the asbestos runoff after the fire. No dollar figures are available for that effort.

FIRE INVESTIGATION

Initial

Bremerton fire and police officials were immediately assisted by a team of regional fire investigators according to the mutual aid plan. The initial investigators were from Kitsap County Fire Marshal's Office. (A Regional Arson Task force pools the resources of fire and police investigators from jurisdictions in the area providing support in the case of large-scale incidents.)

Assistance was requested immediately from the Federal Bureau of Alcohol, Tobacco, and Firearms from the Seattle office. This group was requested because of the magnitude of the interviewing task, the initial fear of a much higher death roll, and the possibility of a criminal act causing the fire.

The ATF Task Force lead the investigation of cause and origin under the authority of the fire chief in a unified command setting. ATF responded with a mobile command and equipment unit and set up at the command post on Sheridan Road. They also established a command post on Sheridan Road. They also established a command presence at the BFD Fire Station 3.

The immediate tasks were to develop and implement a plan for the recovery of the remains of missing persons, conduct critical interviews, and complete a detailed cause and origin investigation.

The recovery involved a canine search and rescue team and the coroner's office. Interviews were conducted with first responding fire units, building occupants, management, and the complex manager. Other persons of interest were also interviewed.

The cause and origin team used photography, heavy equipment operation, evidence collection equipment, and an accelerant detection canine.

Arson was ruled out as the cause of this fire. Combustibles placed too close to the baseboard heater are suspected as having started the fire in Apartment 316.

Assisting Organizations in Investigations

In addition to the BFD and BATF, organizations assisting in the investigation were:

> King County Search and Rescue Team
>
> King County Fire Investigations Unit
>
> Pierce County Fire Marshal's Office
>
> Kitsap County Fire Marshal's Office
>
> Clark County Fire Marshal
>
> Mason County Fire Marshal

CRITIQUES

The Bremerton Fire Department specifies in its Standard Operating Procedures very structured policies and procedures to critique major incidents. The procedures assign duties and responsibilities for implementing the critique sector and establish guidelines for critique levels and time frames. The Kona Village fire definitely qualified for multi-level critiques. A summary of those critiques is listed below; the findings are summarized per specialty.

Bremerton Fire Department

- On Saturday, December 6, 1997, the BFD held a critique of the fire that occurred on November 13, 1997. The results of that critique were published in a BFD Safety Bulletin. The issues and lessons learned from this critique are consistent with the finding of this Technical Report.

Some additional issues were discussed:

- Additional safety officers are needed for incidents of this size and complexity.

- Division and sector officers, as a practicality, must request additional resources as the need for those resources become apparent.

- When tasks are completed or difficulties are encountered, the teams need to inform the group leaders. If resources are needed to complete the task, group leaders or division officers should not hesitate to request them.

- Command and company officers need to be mindful of the danger of asbestos in buildings and to take steps to recognize and mitigate the hazard early in the incident.

- All radio communications were on one channel and this flow of messages generally worked well. However, face-to-face communication between various functions and officers worked much better than did radio communication.

- Structural engineers are required to evaluate damaged structures.

- The use of spray paint to identify the sides of the building and identify rooms searched worked very well.

- Site security is vital for the start of the incident, and through the investigation and recovery phases.

- The rehabilitation unit is vital in extended incidents such as this.

- More aerial devices were required earlier in the incident, because no elevated streams were available to confine fire spread.

- Mutual aid is invaluable, as it directly demonstrates the need for standardization in training and operations.

- Fire departments need to actively promote fire alarm and sprinkler systems in all large apartment complexes.

Support/Services

Due to the size and complexity of the Kona Village fire, a separate critique was held to focus on the non-suppression support services. Many affected organizations attended and their comments and recommendations are summarized by organization:

Central Communications (CenCom)

- Evolution of ICS requires that dispatchers become more active as resource managers.

- Fire operations representatives were needed at CenCom for technical assistance earlier in the incident.

- A PIO representative was needed to field media questions.

- CenCom should be notified when the shipyard fire units are dispatched.

- Problems with run cards and backup have been encountered.

Bremerton Police

- The response went well from their perspective.

- Displaced victims needed to be directed in order to evacuate to same the place.

- Crowd control on the scene and at the hospital depleted all police resources.

- Emergency equipment (such as barricades) is needed.

Kitsap County Sheriff

- The mobile command post was well utilized.

- While the fire was within city limits, much of the command post and support activity were on county property.

Washington State Patrol

- The Washington State Patrol was not needed, however, the extent of their involvement was not clearly communicated.

Bremerton Public Works

- Staff and support vehicles were parked in the way of hydrants and hoselines.

- Hose bridges were needed for supply vehicles in order to have access to equipment for refueling.

- The water department needs to be notified sooner in order to activate supplementary pumps (1.7 million gallons of Bremerton water was used).

- Water system maps need to be carried on command vehicles.

- The development and implementation of a method for supplying fuel to the scene of large operations is needed.

- All equipment needs to be marked. For example, radios should be marked with department and unit members.

- More radio frequencies are needed, with one common frequency for support.

- Hardwired phones, faxes, copiers, printers, televisions, and computers are needed at the command center.

Kitsap County Department of Emergency Management

- A victim advocate is needed.

- An accurate victim list is needed.

- There were problems with victim confidentiality (occupancy list was faxed to an attorney).

- A procedure for accessing personal belongings left at scene is needed.

- The creation of a city phone number with voice mail capacity will facilitate the storage of lengthy amounts of information.

Public Information Task Force (PIO)

- The PIO was needed and well utilized.

- Other entities need PIO's, CenCom, hospitals, and public works administrations.

- Information to the media; consistent facts, need to be coordinated.

- Media representatives should be sent to the PIOs.

- It is crucial that the PIOs have one media contact number.

- A PIO must be present for media interviews with the chief.

- The PIO Task Force should be placed on a run card and called earlier.

- Job descriptions for all PIO members are needed.

- PIO members are required to train and drill as a group, as outlined in the job descriptions.

- The PIO should be present at all command staff meetings.

- The details surrounding the event must be reviewed in their entirety before the PIO talks to the media.

- Integrity of crime-scene-training is needed for PIOs.

- The equipment that is needed includes: cell phones, pagers, batteries, and chargers.

- The PIO must contact the coroner's office in order to verify the number of the deceased.

Coroner's Office

- The morgue permits no more than 10 corpses at one time.

- A prearranged refrigerated truck rental agreement should be in place for deployment as needed.

- An emergency operations plan is needed for the coroner as part of the unified command resources.

- Mutual aid is needed with morgues in other jurisdictions.

- Consider assigning the coroner with investigators.

American Red Cross

- A dedicated phone was needed for use by the Red Cross.

- Crowd control was needed at St. Luke's Church.

- There were problems with cash donations; accountability and security.
- Victims need to be debriefed.

Pastor (St. Luke's Church, evacuation and relief center)

- Prescription medication for victims was a problem because their supply of medications was left behind.
- Crowd control was needed at church.

Department of Community Development

- The plans of the building were needed at the command post.
- Disposition of building remains must be clearly defined; demolition authority, debris removal, and asbestos disposal.
- A policy is needed for calls being made by attorneys to the scene.

Harrison Memorial Hospital

- The hospital needs one central point of communication for incoming reports from the scene, and a contact list for emergency services.
- A log book is needed for documentation of emergency activities.

Investigations:

- Structurally questionable areas should be carefully evaluated for need of shoring.
- Precautions for asbestos are needed.
- Structural engineers are needed before the investigation begins.
- Film runners are needed.
- Support personnel are needed for interviews, time lines, and documentation.
- Administrative supplies (boxes, file folders, computer) are needed at the scene.

Olympic Ambulance

- Though Olympic units were available to help, in some cases, no ambulance was available at the site and victims had to drive themselves to the hospitals.

Fire Administrative Office

- Accurate and timely information is needed because the office received many calls.

LESSONS LEARNED/REINFORCED

1. The use of small 1-3/4-inch handlines was not effective in the exterior operation. While small handlines are more mobile and require less staffing, fire intensity drastically reduced the impact of the small handlines.

2. While variable-pattern fog nozzles are well suited for interior and close-in operations, they were not effective for the range and penetration needed at this fire. Smooth-bore tips were substituted with a marked increase in stream effectiveness.

3. The portable monitor nozzles were used to great advantage at the Kona Village fire. They were rapidly deployed and placed in service quickly when the decision was made to use them. The speed at which they were placed in service reflects a high level of training and competence with these appliances.

4. A quickly deployed aerial master stream in the vicinity of the SW corner would have materially reduced the spread of the free burning fire. A mobile elevated master stream would have been able to knock down the fire that spread laterally in two directions.

5. The staffing level on the initial alarm precluded simultaneous rescue and suppression operations. Forces made the correct decision to concentrate on rescue, with the resultant increase in fire spread.

6. The lack of a sustained water supply in the early stages of the fire placed firefighters in jeopardy and delayed extinguishing efforts. The ground ladders were deployed with rapidity and skill in the rear of the structure, and the manner of their placement reflected a high level of training and competence. Lives were saved because firefighters were able to rescue people who were trapped on the west and south sides.

7. Staffing levels of fire and rescue forces should be a reflection of the code requirements. Unless building codes are written to prevent rapidly advancing fires in residential structures such as Kona Village, fire staffing and operations should be geared accordingly.

8. The lack of readily accessible hydrants hampered firefighters in quickly establishing the water flow required by this fire. A total of three hydrants were within reasonable distance but fences obstructed two of these and curbs required labor-intensive hand lays.

9. The lack of a sprinkler system allowed this fire to gain a lethal headway in a short time.

10. Firefighters were unable to stop the fire in time to prevent the spread of heat, smoke and gases in the walls, attic, and courtyard walkways.

11. Spacing of parking was insufficient in the exterior lot to permit the fire units to operate. Firefighters were hampered by parked cars, residents leaving in their vehicles, and the radiant heat of the fire.

12. Access to the interior courtyard was available only from the breezeways and hampered rescue and suppression efforts.

13. City and fire officials should thoroughly document their observations and recommendations for fire protection improvements. While they may lack the statutory authority to compel fire protection improvement, reference can and should be made to the standard of care for such occupancies in other areas or under current codes. *Neither the authority having jurisdiction, nor the building owner are released from liability from fire protection improvement, even if the code does not require such measures.*

APPENDIX A

Diagrams – Kona Village Complex

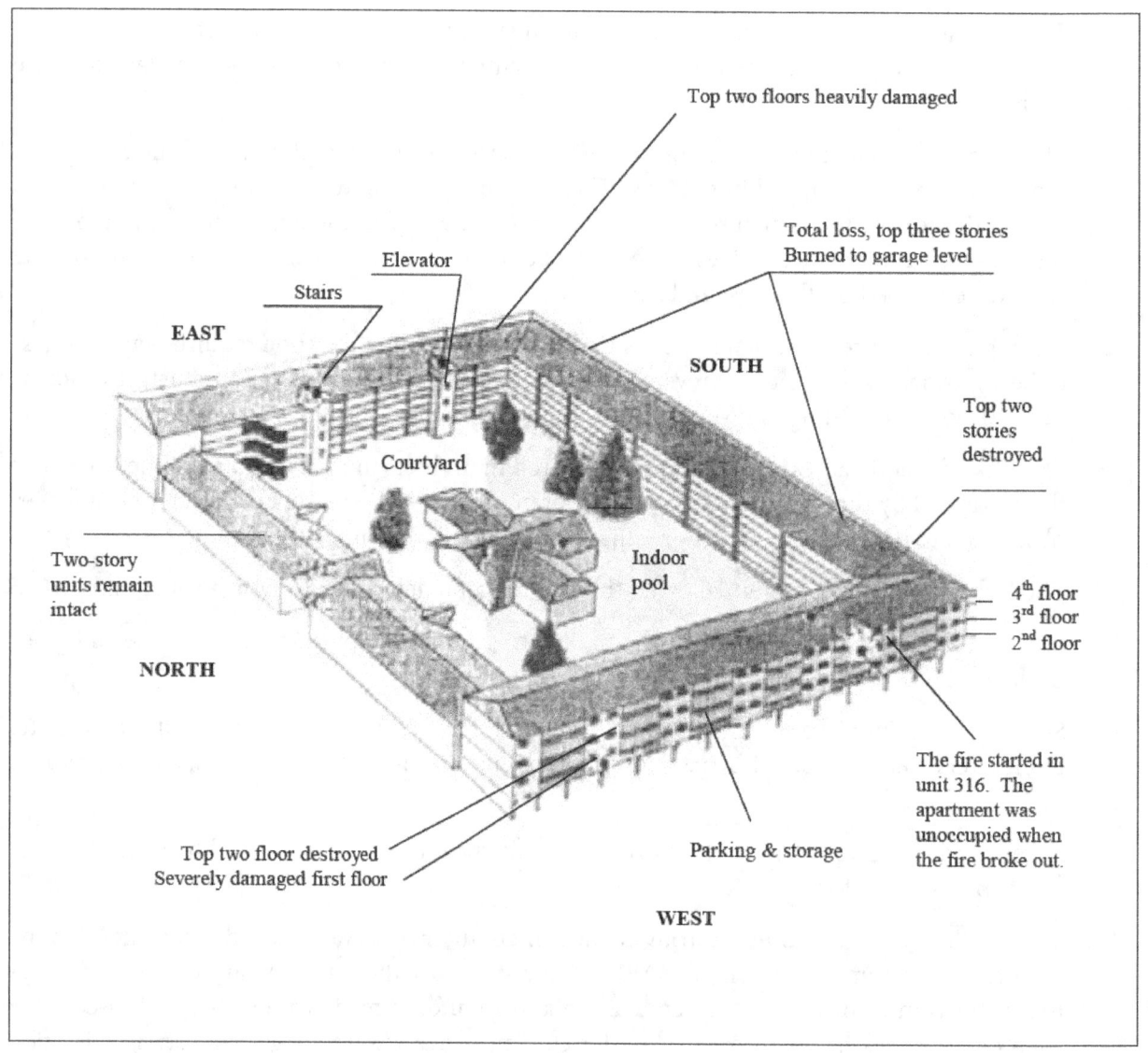

Diagram 1. Kona Village Complex

Appendix A (continued)

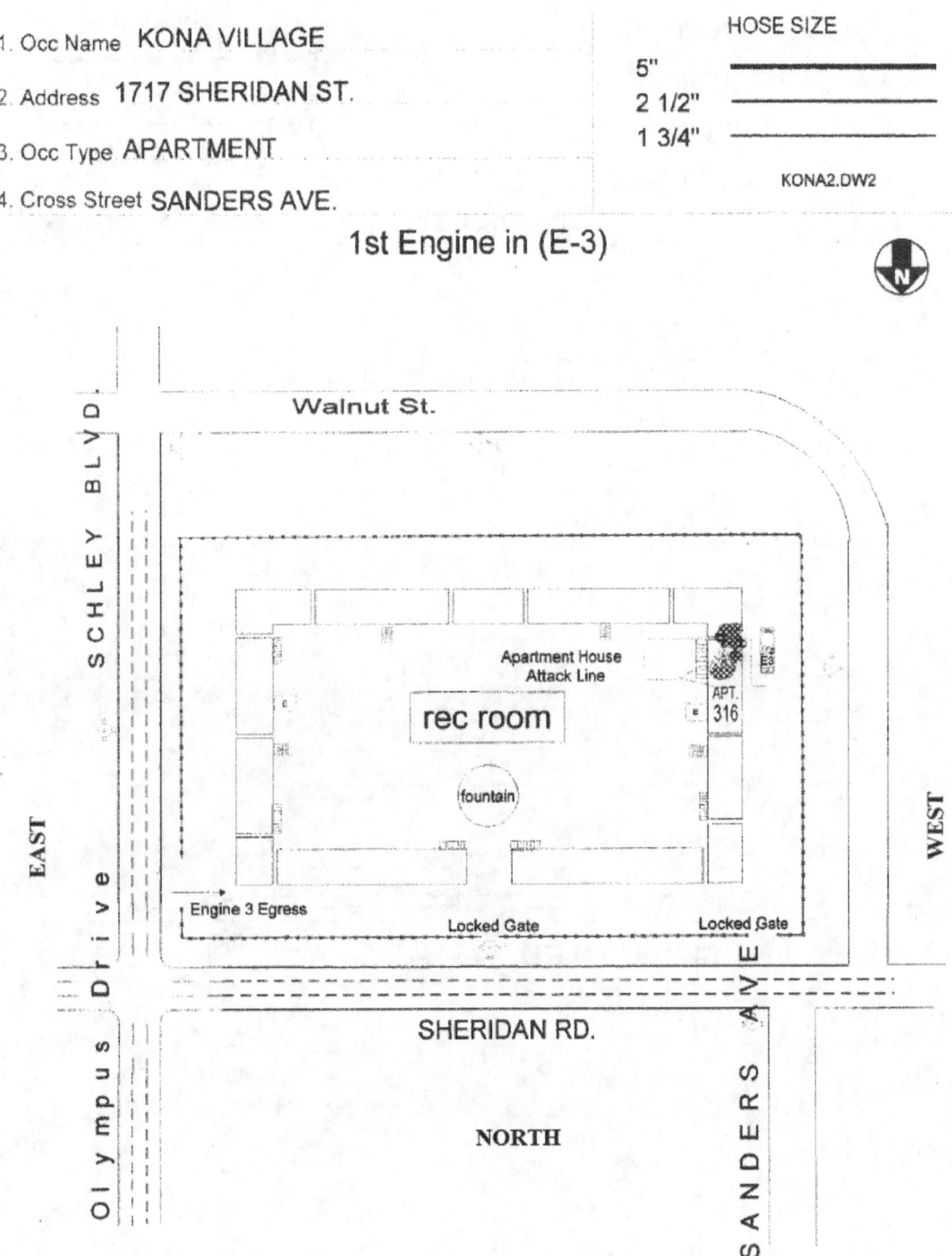

1. Occ Name KONA VILLAGE

2. Address 1717 SHERIDAN ST.

3. Occ Type APARTMENT

4. Cross Street SANDERS AVE.

HOSE SIZE

5"

2 1/2"

1 3/4"

KONA2.DW2

1st Engine in (E-3)

N

Walnut St.

SCHLEY BLVD.

Apartment House
Attack Line

rec room

APT.
316

E

fountain

EAST

Olympus Drive

Engine 3 Egress

Locked Gate   Locked Gate

WEST

SHERIDAN RD.

SANDERS AVE.

NORTH

Diagram 2. Fire origin and first-in engine location

Appendix A (continued)

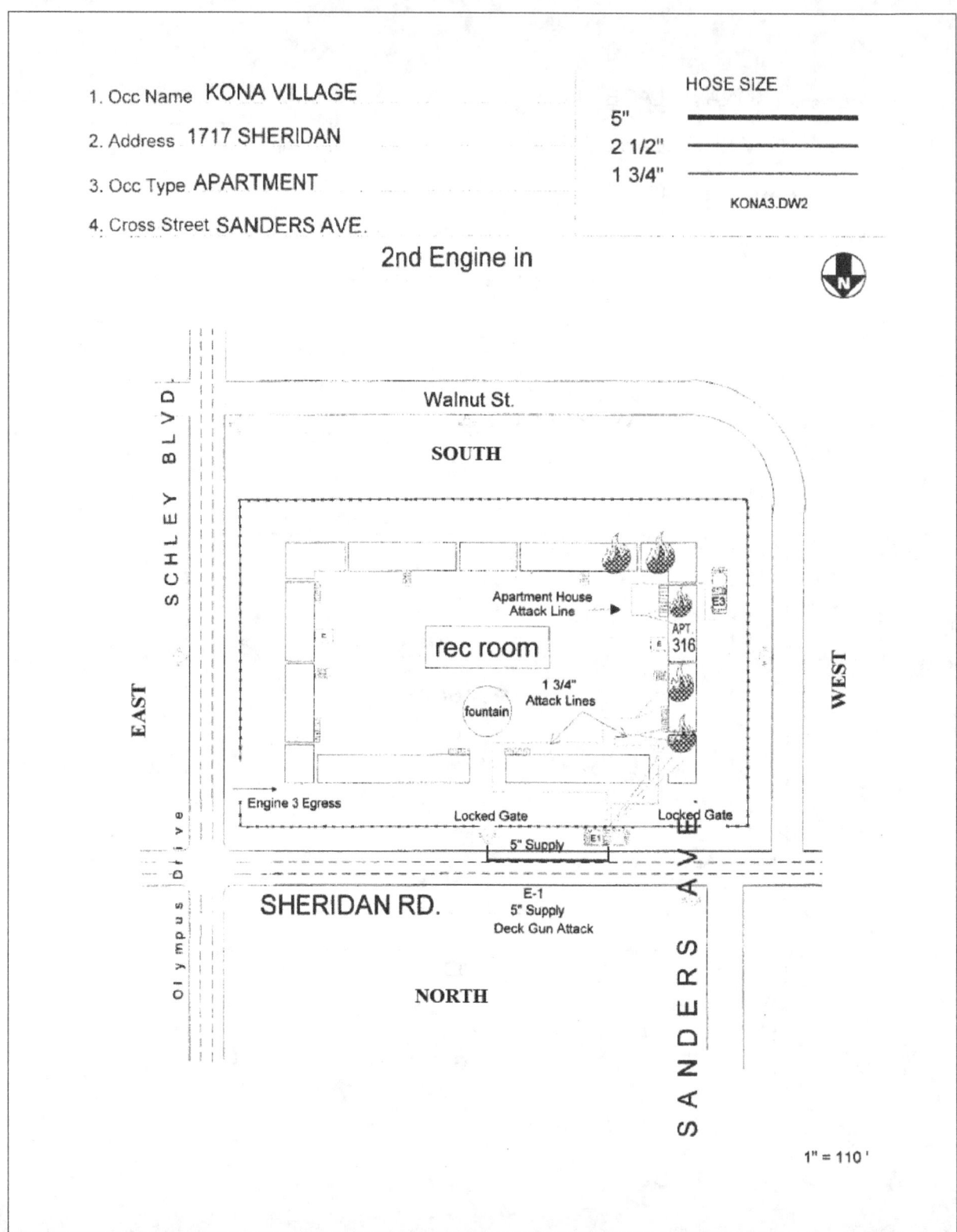

Diagram 3. Position of second engine and early fire travel

Appendix A (continued)

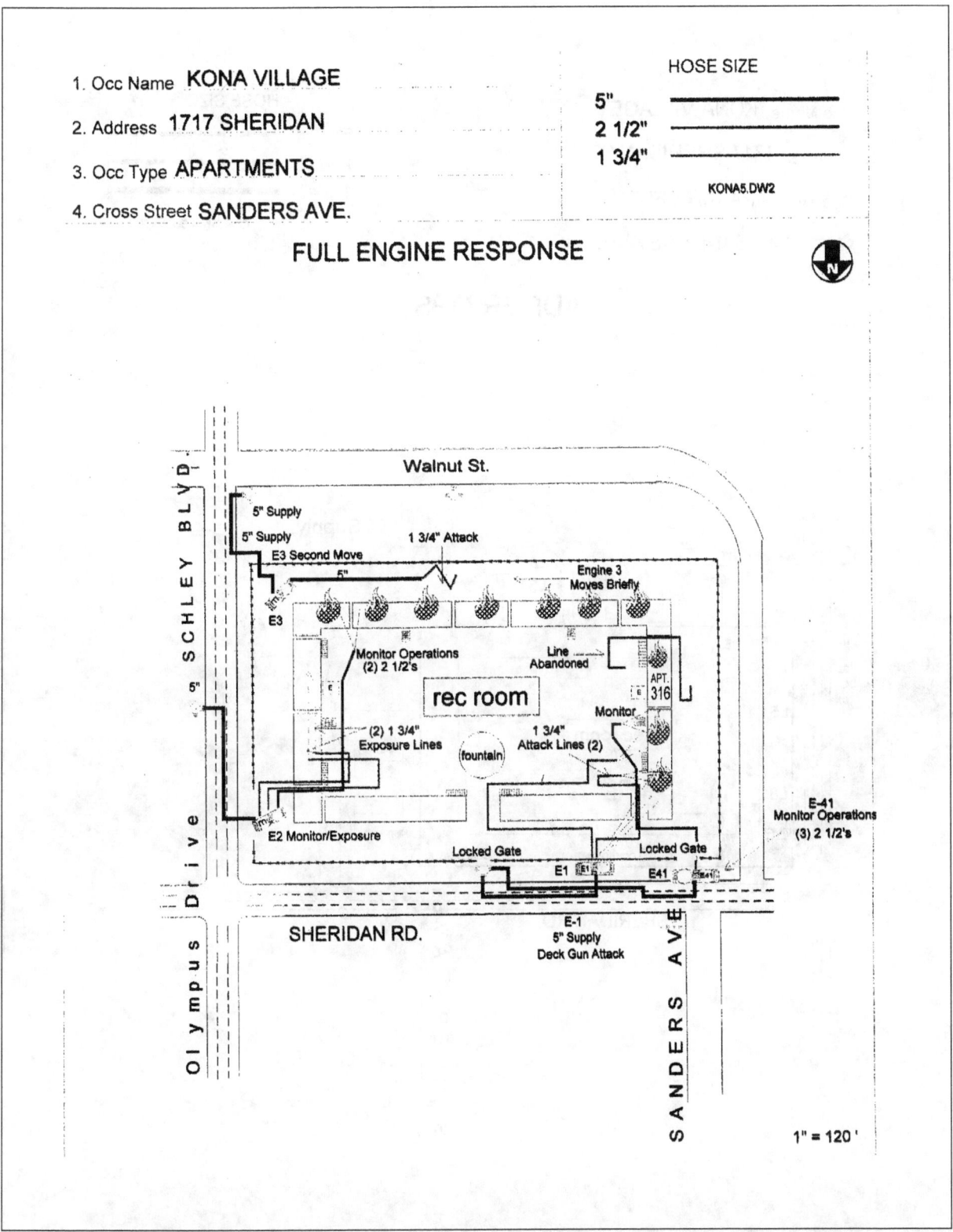

1. Occ Name **KONA VILLAGE**

2. Address **1717 SHERIDAN**

3. Occ Type **APARTMENTS**

4. Cross Street **SANDERS AVE.**

HOSE SIZE

5"

2 1/2"

1 3/4"

KONA5.DW2

FULL ENGINE RESPONSE

Walnut St.

5" Supply
5" Supply
E3 Second Move
1 3/4" Attack
5"
Engine 3
Moves Briefly

E3

Monitor Operations
(2) 2 1/2's
Line
Abandoned

APT.
316

rec room
Monitor

(2) 1 3/4"
Exposure Lines
1 3/4"
Attack Lines (2)

fountain

5"

E-41
Monitor Operations
(3) 2 1/2's

E2 Monitor/Exposure
Locked Gate
Locked Gate

E1 E41

SCHLEY BLVD.

Olympus Drive

SHERIDAN RD.

E-1
5" Supply
Deck Gun Attack

SANDERS AVE.

1" = 120'

Diagram 4. Third phase of operations; cut-off fire spread, attack and extinguish

Appendix A (continued)

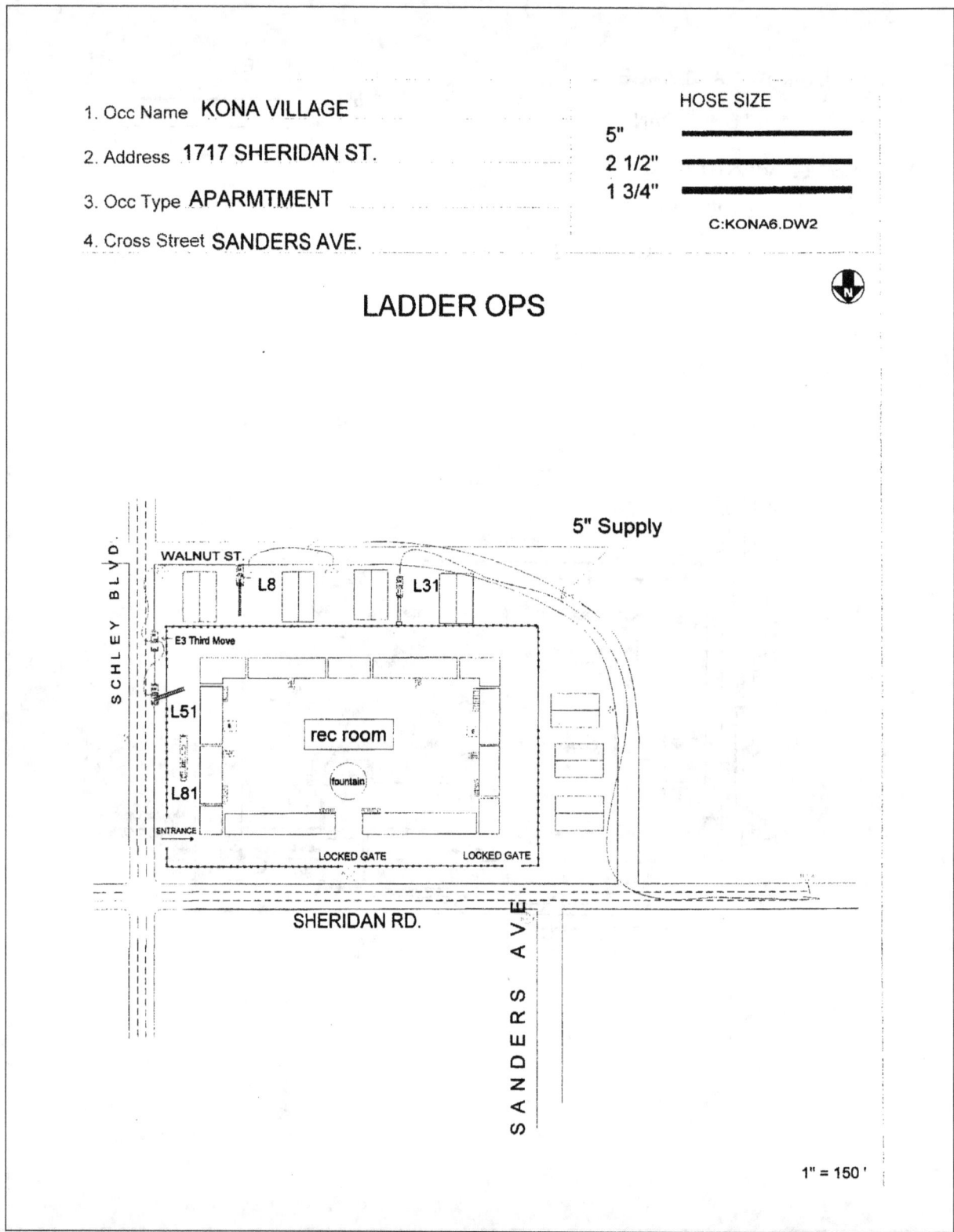

Diagram 5. Ladder company position

APPENDIX B

Photographs

Photo 1. Extreme fire conditions at the southeast corner

Photo 2. Interior operations in courtyard, west wing

Appendix B (continued)

Photo 3. The view from the interior courtyard toward the SW corner.
The X approximates the apartment of origin. Note the
hose cabinets and piping still standing

Photo 4. An elevated view of the west wing looking in a NW direction.
The apartment of origin is at the center left

Appendix B (continued)

Photo 5. An elevated view of the south wing toward the east wing.
The fire was stopped at the SE corner

Photo 6. The view from the interior courtyard looking toward the NW corner

Appendix B (continued)

Photo 7. A Kona Village resident just rescued from upper floor by first arriving firefighters anxiously awaits the fate of her husband. Flames in background illustrate conditions on arrival of the fire department

Appendix B (continued)

Photo 8. Rescue of elderly tenant, spouse of woman in Photo 7, underscores problems faced by first arriving units of Bremerton Fire Department and Puget Sound Naval Shipyard Fire Department (Sun Photos by Larry Stegall)

APPENDIX C

Agency Participation: Kona Village Fire, 11-13-97

Agency Participation

Bremerton Fire Department

Puget Sound Naval Shipyard

Kitsap County Fire District #1

Kitsap County Fire District #15

Kitsap County Fire District #12

Kitsap County Fire District #7

Kitsap County Fire District #10

Kitsap County Fire District #18/

Poulsbo Fire Department

Mason County Fire District #2

Mason County Fire District #5

Mason County Fire District #3

Jefferson County Fire District #4

Mason County Fire Marshal's Office

Kitsap County Fire Marshal's Office

Clark County Fire Marshal's Office

Pierce County Fire Marshal's Office

Regional Fire One Marshal's Office

Regional Fire Two Marshal's Office

Bremerton City Council Office

Bremerton Mayor's Office

Bureau of Alcohol, Tobacco and Firearms

National Fire Protection Association

IAFC – Operation Life Safety

Central Communications – CenCom

Department of Emergency Management

Kitsap Transit

Kitsap County Sheriff's Office

Washington State Patrol

West Sound American Red Cross

Saint Luke's Church

Puget Sound Energy

Bremerton Public Department

Green Springs, Inc.

Honey Buckets

Annie's Restaurant

Kitsap County Coroner's Office Olympic Ambulance

Bremerton Ambulance

Bremerton Department of Public Works

  Electronic Division

  Water Division

  Street Division

  Engineering Division

  Transportation Division

Bremerton Department of Community Development

Bremerton City Attorney

United States Fire Administration

Bremerton Parks and Recreation

Total of 44 Agencies

Appendix C (continued)

News Media Agencies:

The Bremerton Sun

KIRO TV & Radio (7)

Central Kitsap Reporter

KSTW (11)

KOMO TV & Radio (4) Seattle Pl

KING TV & Radio (5) Seattle Times

Agencies Offering Assistance:

Bainbridge Island Fire Department

Auburn Fire Department

Seattle Fire Department

Enumclaw Fire Department

Tacoma Fire Department

Willow Retirement Center

www.ingramcontent.com/pod-product-compliance
Lightning Source LLC
Chambersburg PA
CBHW081242170526
45165CB00009B/3159